Genre Nonfiction

Essential Question

How do volcanoes and earthquakes change the land?

How Earthquakes and Volcanoes Shape Earth

by Barbara Collier

Chapter 1

Shaking Things Up

Roar! An **earthquake** shakes the ground. Crash! A tree falls down. Whoosh! A **volcano** explodes. **Lava** flows out. Lava is very hot liquid rock. It destroys everything in its path.

Volcanoes and earthquakes are **natural hazards**. A natural hazard is caused by forces in nature. It can harm people. A hurricane is a natural hazard. A tornado is a natural hazard, too.

Suppose people could stop all natural hazards. Would that be a good thing? It might not be. Earthquakes and volcanoes can cause harm. Yet they can be helpful, too. Earthquakes and volcanoes helped to shape Earth.

---

suppose: to imagine, or think to be true
shape: to make, or give a form to

There was no atmosphere when Earth was a young planet. Volcanoes **erupted** for millions of years. Lava and gases from inside Earth poured out. Those gases became part of the atmosphere. Some formed rain. It fell and became the first ocean.

About 250 million years ago, Earth had just one big piece of land. The ocean was around the land. Then Earth shook and things crashed. Lava poured out of the ground. The earthquakes and volcanoes helped break the land into pieces. Now, millions of years later, there are different land areas. They are **continents**.

volcano

Chapter 2

Our Puzzle Planet

The crust is Earth's outer layer of solid rock. The mantle is the layer below the crust. Very large pieces form the crust and upper mantle. The pieces are called **tectonic plates**. They fit together like pieces of a jigsaw puzzle. They move over a layer of hot, almost-melted rock that flows.

Earth's continents cover some parts of these plates. The oceans cover the other parts.

---

jigsaw puzzle: cut up pieces of a picture that must be put back together

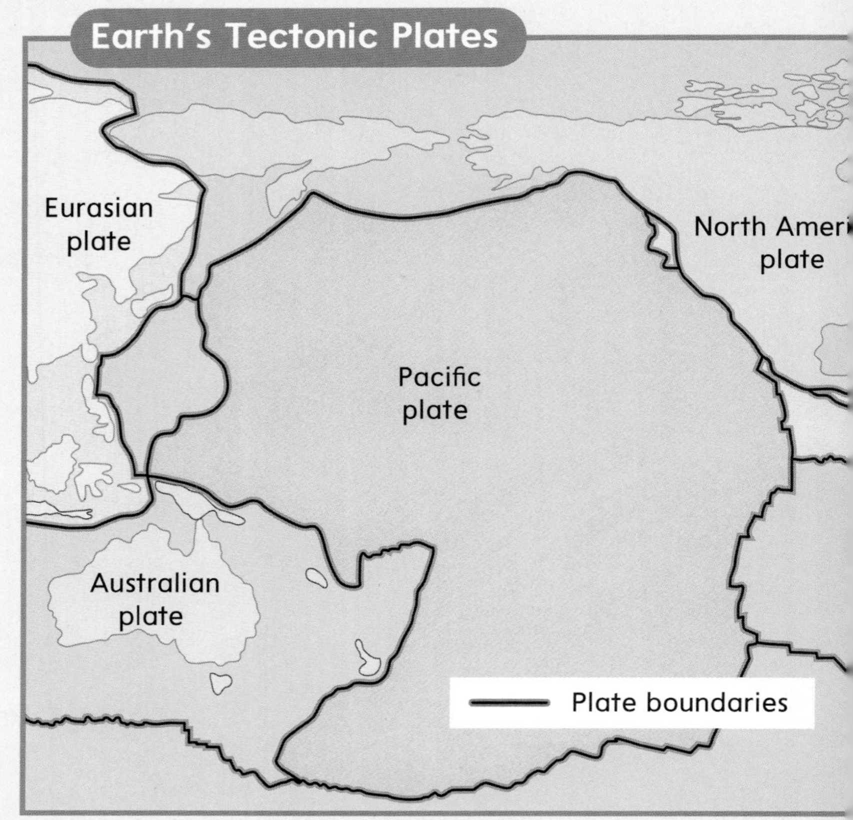

The plates move very slowly. Usually, you cannot tell they move.

Think about a lake filled with big floating rafts. Some rafts bump into each other. Other rafts float apart. Others slide past each other. Earth's plates move like the rafts. They move together, apart, and past each other. Two plates meet at a **plate boundary**. Many earthquakes happen near plate boundaries. Most volcanoes form near plate boundaries.

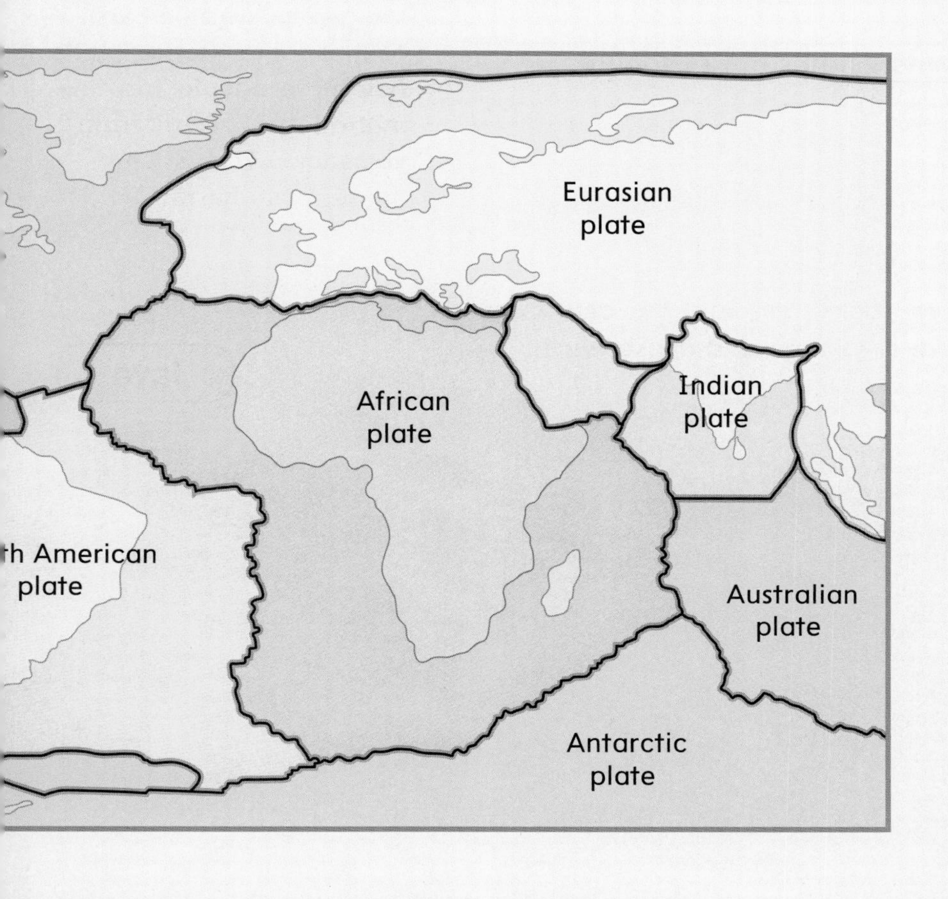

Chapter 3

Volcanoes

Two tectonic plates can come together at a plate boundary. One may slide under the other. It sinks into the hot mantle. Some of this plate melts. The melted rock moves up through cracks in the rock above. At the surface, a hole can form. Then hot lava bursts out of the ground. A volcano forms!

When a volcano erupts, rocks, ash, and hot gases burst out with the lava. This can destroy everything nearby.

bursts: comes out suddenly

Another Look

In some places, people use water heated by melted rock inside Earth to make electricity. That's helpful to people. Yet this melted rock erupts from volcanoes. Then it can cause much harm.

Lava, rocks, ash, and gas can erupt from a volcano and cause harm.

The city of Pompeii was destroyed when Mount Vesuvius erupted.

Almost 2,000 years ago, Pompeii was a town in Italy. It was near a beautiful mountain called Mount Vesuvius. The mountain was really a volcano. It had not erupted for hundreds of years.

Suddenly, the top of the mountain blew! People tried to run away. Rocks and ash buried the town and many people. Pompeii is an example of how volcanoes can cause great harm.

Volcanoes can destroy things. They can also build up Earth's surface! After lava erupts onto the surface, it cools and hardens into new rock. This happens again and again. A new layer of rock forms each time. This rock builds up Earth's surface.

The rock layers slowly build up and mountains form. This happens on continents and under the ocean. An island forms when enough underwater rock builds up to reach above the ocean surface. The islands of Hawaii formed this way.

A volcano that erupts pushes out ash, too. This ash can harm living things. It can also help by adding minerals to soil. Minerals are nutrients that living things need. When ash mixes with soil, it makes the soil better for growing plants.

Chapter 4

Earthquakes

One spring day in 1960, the ground shook in the town of Valdivia, Chile. People rushed out of their houses. An earthquake knocked down almost all the buildings. It was the biggest earthquake recorded.

A street in Valdivia, Chile, after the 1960 earthquake

Earthquakes are similar to volcanoes in one way. Tectonic plates that move cause most earthquakes. Cracks called **faults** form in Earth's crust as plates move and change. The rocks on either side of a fault can move. When they move suddenly, an earthquake happens. The crust shakes with a little or a lot of energy. The energy spreads out through Earth as waves. The ground shakes a little bit in a small earthquake. In a big earthquake, houses, cars, and even boats in the ocean move. Buildings and roads are damaged or destroyed.

Earthquakes can destroy things. Yet when two plates push together, they push up land. Faults form. Rock on one side of a fault gets pushed up over rock on the other side. This can form mountains. Parts of the Rocky Mountains formed this way.

Mountains can form when rocks get pushed up along a fault.

Chapter 5

Staying Safe

People cannot stop natural hazards. They can find ways to stay safe if they do happen.

Standing Up to an Earthquake

Most harm to people from earthquakes happens when buildings or bridges fall down. Builders now use strong steel frames and other features. These features make buildings safer during earthquakes.

Living with Volcanoes and Faults

Scientists use special tools to study changes in the ground near volcanoes and faults. These tools help tell if a volcano might erupt or an earthquake might happen.

Learning from Natural Hazards

Natural hazards can be dangerous. However, they also help shape Earth's surface. Scientists study them so they can learn about Earth.

This new bridge was built to withstand earthquakes.

Summarize

Use details from *How Earthquakes and Volcanoes Shape Earth* to summarize the selection. Your graphic organizer may help you.

Cause → Effect
→
→
→
→

Text Evidence

1. How are earthquakes and volcanoes similar? How are they different?

2. Read the book again with a partner. Explain what causes earthquakes and volcanoes and what happens as a result. CAUSE AND EFFECT

3. What is the meaning of the word *break* on page 3? What is another meaning for the word *break?* What clues in the text show you which meaning to use on page 3? HOMOGRAPHS

4. Work with a small group. Summarize the "good" effects of earthquakes and volcanoes. Share your summary with another group. WRITE ABOUT READING

Compare Texts

Read how Jenna and her cousin watch an active volcano in Hawaii. Learn how volcanoes work.

Lava-Flow

The buzzer of the alarm clock rang. Jenna blinked twice. It was 5:30. Time to get up!

Jenna was still tired. She also felt excited. Today was the day that her cousin, Mosi, was going to take her to visit Kilauea. It was a volcano that was erupting in Hawaii. Ever since she arrived to spend two weeks with her aunt and uncle, she had hoped to see a volcano.

Jenna and her cousin left early that morning. Mosi drove across the island roads.

Mosi said, "See those fields over there." They passed several farms. "Those orange trees are growing on the old lava beds."

Mosi explained that the volcanoes on the island could cause problems. Yet crops grew well on the old volcanic soil. At last, Mosi turned down a road and parked his car.

Mosi said, "Stay close behind me so we don't get too close to the lava." Jenna knew that moving lava could surprise visitors if they were not careful.

Mosi and Jenna found a spot to see the lava move slowly over the rocks toward the ocean. They watched as the stream of lava flowed off the rock and into the Pacific Ocean. A cloud of steam came up from the water.

Jenna said, "It's so amazing!" She watched as the lava cooled in the ocean. Right before her eyes was a volcano slowly adding to the history of Hawaii.

Make Connections

What does Jenna learn about volcanoes?

TEXT TO TEXT

Glossary

continent *(KON-tuh-nuhnt)* one of several large land areas on Earth *(page 3)*

earthquake *(URTH-kwayk)* a sudden shaking of the rock that makes up Earth's crust *(page 2)*

erupted *(i-RUHPT)* burst or poured out *(page 3)*

fault *(FAWLT)* a crack in Earth's crust along which movement has taken place *(page 10)*

lava *(LAH-vuh)* hot, melted rock that reaches Earth's surface *(page 2)*

natural hazard *(NACH-uhr-uhl dih-ZAStuhr)* an event caused by natural forces on Earth that can harm people *(page 2)*

plate boundary *(PLAYT BOUN-duh-ree)* the place where two tectonic plates meet *(page 5)*

tectonic plate *(tek-TAHN-ik PLAYT)* a huge piece of Earth's solid crust and upper mantle that moves around the surface *(page 4)*

volcano *(vol-KAY-noh)* an opening in Earth's crust through which lava may flow *(page 2)*

Index

Purpose To model an earthquake

Procedure

tep 1 Wrap board erasers in plastic wrap and cover a small table with newspaper.

tep 2 Place the erasers on the desk close together.

tep 3 Sprinkle soil over the erasers.

tep 4 Rock the desk back and forth.

Conclusion What do the erasers represent? What does the soil represent? What happened to the erasers and soil when you shook the table? How does this model an earthquake?

Lexile 630

www.mheonline.com/inspire-science

978-0-02-137378-9
MHID 0-02-137378-7
EAN
9 780021 373789
99701
4

Mc
Graw
Hill
Education

onfiction

Maglev Trains

by Andrew Seear

PAIRED READ Smooth as Air

STRATEGIES & SKILLS

Comprehension

Strategy: Summarize

Skill: Compare and Contrast

Vocabulary Strategy

Homographs

Vocabulary

electromagnetic, environment, friction, guideway, levitation, magnet, magnetic natural resource, pollute, vibrate

Content Standards

Science

Physical Science

Word Count: 1518**

Photography Credit: Cover ©Viewstock/Corbis

**The total word count is based on words in complete sentences found in the running text, sidebars, headings, and captions. Numerals and words in phrases that comprise labels, diagrams, and headings are not included.

ConnectED.mcgraw-hill.com

Send all inquiries to:
McGraw-Hill Education
8787 Orion Place
Columbus, OH 43240

ISBN: 978-0-02-135918-9
MHID: 0-02-135918-0

Printed in China.

3 4 5 6 DSS 20 19 18 17

A